YOUR KNOWLEDGE HAS VALUE

AF152021

- We will publish your bachelor's and master's thesis, essays and papers

- Your own eBook and book - sold worldwide in all relevant shops

- Earn money with each sale

Upload your text at www.GRIN.com and publish for free

Dimitrios Kamsaris

The Brain-Net of Communication

Persuasion in Action

GRIN Verlag

Bibliografische Information der Deutschen Nationalbibliothek:

Die Deutsche Bibliothek verzeichnet diese Publikation in der Deutschen National-
bibliografie; detaillierte bibliografische Daten sind im Internet über http://dnb.d-
nb.de/ abrufbar.

Imprint:

Copyright © 2003 GRIN Verlag GmbH
Druck und Bindung: Books on Demand GmbH, Norderstedt Germany
ISBN: 978-3-656-75537-1

This book at GRIN:

http://www.grin.com/en/e-book/280897/the-brain-net-of-communication

GRIN - Your knowledge has value

Der GRIN Verlag publiziert seit 1998 wissenschaftliche Arbeiten von Studenten, Hochschullehrern und anderen Akademikern als eBook und gedrucktes Buch. Die Verlagswebsite www.grin.com ist die ideale Plattform zur Veröffentlichung von Hausarbeiten, Abschlussarbeiten, wissenschaftlichen Aufsätzen, Dissertationen und Fachbüchern.

Visit us on the internet:

http://www.grin.com/

http://www.facebook.com/grincom

http://www.twitter.com/grin_com

The Brain-Net of Communication
"Persuasion in Action"

Author:

Prof. Dimitrios P. Kamsaris

Affiliation: Bilston Community College, UK

Abstract

This book aimed at answering the question trying to explore to way in which the perception of the brain influence the communication that is prevailing within human interaction. Further to that, does communication result to conflict or conflict resolution? What is the problem represented to be? What presuppositions underlie the conceptions concerning the communication process? What effects are produced by such representations? How are subjects constituted within the communication?

Keywords:
Communication; Perception; Brain

1. Introduction

Every entity has to continuously get strengthen to sustain a competitive advantage (Barney, 2008), in order to exist (Brown and Kaka, 2003). Further to that one must keep in mind that people are stimulated by two the types of environments, the external environment which includes the market, the customers, the stakeholders that has to be carefully defined and assessed, and the economic, political, social and technological situations, etc. and the internal environment the strengths and the weaknesses etc. (Kelly and Bowles, 2006).

This book aimed at answering the questions, in which way does the perception of the mind influence the communication that is prevailing within human interaction. Further to that does communication result to conflict or conflict resolution? What is the problem represented to be? What presuppositions underlie the conceptions concerning the communication process? What effects are produced by such representations? How are subjects constituted within the communication?

In summary, the view taken in this book is: (a) in every entity a specific forms of communication that prevail and, (c) the forms of communication are combined in a unique pattern of communication.

1.1. Defining the Problem:
Communication Pattern:
- The focus of the book is on the forms of communication and the communication pattern.

1.2. The Research Questions
The overall strategic question to be answered is:
- In which way does the perception influence the communication that is prevailing in an entity?

The secondary questions the present study is answering are:

- Does communication result to conflict or conflict resolution?
- What is the problem represented to be?
- What presuppositions underlie the conceptions concerning the communication process?
- What effects are produced by such representations?
- How are subjects constituted within the communication?

2. Supporting Bibliography

To explore how communication is shaped it is necessary to define the base concepts that will be used in the book research. In this part the different aspects will be introduced and basically defined while the elaboration of the aspects will be completed further during the second part of the book.

2.1. Conceptualizing the Entity

The organization theory is supported by sociological disciplines focusing on the organizational stability (Burke, 2002) and are concerned with open-systems theory derived from the biological sciences field, and organizational psychology based on the management field (Schein, 1985). The open-systems viewpoint concerns the bond developed between the company and the environment it inhabits (Beer, 1980).

There are three schools of thought in institutional theory the historical institutionalism, where procedures, routines, norms and conventions embedded in the company structure, the rational choice institutionalism, where company employees have a fixed set of preferences, behave instrumentally, so as to maximize the attainment of these preferences and the sociological institutionalism , where a company include rules, procedures, norms, symbols, offering a frame of meaning guiding behavior (Hall & Taylor, 1996).

There are differences in communication patterns, preferences for leadership style, different principles of hierarchy, organizational structures, decision styles, and dyadic relationship (Kuchinke, 1999). There are five dimensions that provide a framework for identifying similarities and differences which are the Power Distance, Individualism, Masculinity, Uncertainty Avoidance and Long term Orientation (Hofstede, 1993). It is less risky to communicate with more similar cultural clusters (Gupta, et al, 2002).

2.2.   *Forms of Human Communication*

"Communication," which is etymologically related to both "communion" and "community," comes from the Latin communicare, which means "to make common" (Weekley, 1967) or "to share." Communication is the flow of information in order to share the meaning and each company has its own approach to transmitting the information throughout the entity and may assume different forms such as verbal or written, formal or informal (Bovee & Thill, 1992)

Verbal communication is used to discuss events, ideas and for sharing information and is equally, if not more important as written communication while formal communication refers to the exchange of official information that follows the chain of command, whereas, informal communication uses the unofficial lines of activity (Bovee & Thill, 1992). In order to communicate a reaction by the receiver is required as feedback where, features of context such as "who communicates, why, when, where and how, contribute overt and covert information to the exchanges produced and have to be taken into account for the decoding of the intended message (Sifianou 2001: 2).

Effectiveness of communication depends on the closeness of meaning that the sender and the receiver attribute to the message (Hodgetts & Luthans, 1991). To have an effective argument, the emotion should be recognized by the incentive and should not be supplementary; otherwise, these may seem as a biasing factor, making receivers try to remove the unfitting influence by correcting it (Wegener & Petty, 1997).

2.2.1. Communication Process

When two people communicate they attempt to change the way of thinking, feeling, knowledge or behavior (Argyris, 1977). Communication is interactive where the sender and the receiver are interdependent due to the fact that we depend on the other for a response or the interaction is scattered, through a channel and with the interference of noise and is intentional and conveys meaning in order to bring about change (DeVito, 1986).

6

Sometimes, there is pointless communication intended only to fill silence, weakening from the message to be communicated. The effective communicators should persist much focused on the communicated idea by making sure that their words and actions communicate without any distractions of their message. People communicate through symbols which can take the form of a language, sound, behavior, letter, image that has the meaning of something else and sometimes leads to misunderstanding which requires a correction of the sent message.

The perceptions of the sender are the source of information to be communicated towards the receiver, while the perceptions of the receiver outline the result of the communication. The memory and perception are selective about the acceptable and the information remembered in temporary or long-term memory, so perceptions are not directly correlated to specific events or objects, but are created by the individual system of perception.

2.2.1.1. Message Content

Message is a signal, combination of signals or symbols that serves as a stimulus for a receiver (DeVito, 1986). Content of a message is the substantive information that is being transmitted and the style with which it is being conveyed. Consider the following:

- Persuasive is a carefully reasoned message or one stimulated by emotion? The debate is between reason and emotion. If the receiver is educated and analytical, the rational appeal works better since they are open to reasoned arguments than receivers not familiarized to analytic thinking (Brinol, Petty, Gallardo, &De Marree, 2007).

- Support an idea that is different other's or to an extreme point of view? New ideas produce discomfort, which causes people to change their opinions in response to a new idea. A more direct and bold approach is preferable, although, sometimes an uncomfortable message causing people to lose credibility. People are open to assumptions within their range of acceptability. Greater difference between the new and the sender's established attitude result smaller change in attitude (Brinol, Petty, Gallardo, &De Marree, 2007).

- Message is expressing one side of the argument, or should present the opposing view? The messages content is more persuasive as it is connected with good receiver feelings. Happy people make quick decisions and unhappy make slow decisions, so it would be helpful to make people feel positive. Contrariwise, some messages are persuasive because they appeal negative emotions such as fear and anger, which are strong motivators(Brinol, Petty, Gallardo, &De Marree, 2007).

- Does the order of speaking give the speaker any advantage? There are circumstances when people weaken their argument and confusing the audience. Primacy and recency effect designates that the person or argument making the first impression has the most influence on the decision outcome.

  o The primacy effect states that in the short term, the first argument is the most influential, all things being equal. People think in duality terms as they view argument on a continuum with the truth lying somewhere between the two extremes.

  o The recency effect state that after some time, the primacy effect fades from lively thoughts and feelings. So, forgetting has the effect of creating a more powerful recency effect when sufficient time separates the two messages and people must decide soon after the second argument (Brinol, Petty, Gallardo, &De Marree, 2007).

2.2.1.2. Message Is Conveyance

Communication channel refer to the way a communication is delivered such as face-to-face, in writing, telephone, the medium, in other words. People developed more confidence in their thoughts to a message and relied on them more if, following the message, they learned that the source was of high credibility (Tormala, Brin˜ ol, & Petty, 2006), were made to feel powerful (Brin˜ ol, Petty, Valle, Rucker, & Becerra, 2007), experienced fluency in thought generation (Tormala, Petty, & Brin˜ ol, 2002), or affirmed an important value (Brinol, Petty, Gallardo, &De Marree, 2007).

Tips for conveyance:

- The sender will pay attention to the message

- The sender will understand it

- The sender will believe it

- The sender will remember it.

- The sender will behave as expected

- One medium of communication seem not to be effective enough (Brinol, Petty, & Tormala, 2004),

2.2.1.3. The Audience

It is important to be accepting of the characteristics the audience acquires before presenting a persuasive argument, which are listed below:

- Attitudes about specific issues are significantly or temperately related to decisions.

- Life experiences relevant to specific issues shape the listening and mental processing of the receivers. High self-esteem realizes an argument faster, but is often prone to change their opinions because their opinions are held more confidently.

- Personality traits of people influence their perceptions and their decision and if we cannot predict their responses to an argument may improve the effectiveness of communication.

- General values/beliefs and attitudes of audience's thinking about specific issues.

- Influential demographics are not helpful in predicting how they will process but can raise a question or create a hypothesis in a particular issue that can be helpful in forming a composite understanding Brownstein, 2003).

A persuasive attempt is effective when the receiver:

- Knows, likes, believes and respects the sender.
- Already believes in the message.
- Is predisposed to act on behalf of their beliefs.
- Already has a history of acting successfully on behalf of your cause.
- Is further motivated to act by benefits appealing specifically to them.
- Is capable of taking the desired action.
- Has enough time and resources to take the desired action.

2.2.2. Persuasion

Persuasion is a time consuming effort where symbols are used, and include the active involvement of the receiver of the message, where the sender try to transfer people from one state to a next appreciated state that solves the problem best, with messages transmitted through language cultural meanings to change attitudes. (Perloff, 2003)

2.3. Communication and Persuasion

Persuasion is approach beginning with ancient Greeks such as Aristotle and continuing to the Italian Renaissance with Oratorios from (McGuire, 1969). The process of persuasive is activated with basic communication as the deliberate and focused interface between people through the use of verbal and nonverbal symbols in order to motivating someone to act in a certain way. Persuasion is a symbolic process in which people try to convince each other to change their behavior regarding an issue through the transmission of a message (Perloff, 2003). There is the indirect phenomenon of the pre-decisional distortion of information (Russo, Medvec, & Meloy, 1996) and the effect of viewed as being manipulated by others (Calfee & Ringold, 1994),

To practice persuasion, we must understand that other people have desires and beliefs, a mental state that is vulnerable to change, and has a different perspective than they do (Perloff, 2003). As we develop we rely less on coercive social influence attempts than on persuasion, and develop the ability to persuade others more effectively (Kline, 2005). People's commitment to an idea develops with understanding (Calfee & Ringold, 1994),

Coercion is a technique of forcing people to performance contrary to their preferences with the use of threats if people do not act as demanded (Perloff, 2003).It's all a matter of how people perceive things, as when individuals perceive that they have no choice but to comply, the influence attempt is better viewed as coercive (Smith, 2000).

2.3.1. Effects of Persuasion

Communications exert three different persuasive effects (Thompson, 1999):

- Shaping: Attitudes are shaped by correlating pleasurable environments with a person, or idea.
- Reinforcing: Persuasive communications are designed to reinforce a position they already hold.
- Changing: Communications change attitudes. (Thompson, 1999).

People may of the messages more when they are sad state rather than a happy because sadness signals a problem to be solved (Schwarz, Bless, & Bohner, 1991) or because it conveys a sense of uncertainty (Tiedens & Linton, 2001). When the effect of emotion is low, an experienced emotion can bias thoughts about an object (Petty et al., 1993) resulting in positive consequences when people are in a happy state (Desteno, Petty, Wegener & Rucker, 2000, Petty, Fabrigar, & Wegener, 2003). Many variables bias thinking such as own accessible attitudes (Fazio & Williams, 1986), emotions (Petty et al., 1993), the credibility of the source (Chaiken & Maheswaran, 1994), when people are interested and the message (Asch, 1948).

2.3.2. Creating and Changing Perceptions and the Associated Tools

Many tools have been developed to create, maintain, and change perceptions, which are consequent of accepting about the cognitive processes (thinking) and the affective processes (feeling) Chaiken & Maheswaran, 1994). A selection of these tools that are the language, the schemas, and the affect.

Language: Language is a powerful communication tool as it helps organize the thoughts and the information in a systematic way and is used as words, concepts, or style of expression. The words transfer meaning and concepts, and the same words can be articulated or demonstrated in many ways as it is flexible. The concepts are developed as symbols and concepts not as spelling and the formal rules of grammar that is why it is important to be concerned about the concepts and the connotations than the actual words that will follow, if the concepts are right. Nevertheless, sometimes language is seen as a technical construct where the use of particular words that fit patterns rather than focus on concepts and messages that people must hear and see if we are to be persuasive. Through words we can create a mental picture.]

Schemas: Schemas is a term used in psychology describing people's tendency to form pre-conceptions and expectations as a result of previous life experiences. Schema is a cluster of concepts and perceptions and the system used to organize them so as to support us to make rapidly sense of the events occurring. The brains tend to store information in an order so as to make sense and order and to avoid disorder or inconsistencies and schemas are formed predominantly by perceptions and ideas related to us through others. Socialization is an influence on how schemas and stereotypes are formed, maintained, and changed, in which the significant people have an intense influence on our perspectives. Even when the facts are clear, people have a dissimilar form of the same schema to understand what happened, why things happened in the specific way, and what the result should be. Since schemas are pre-existing, we must have dominant messages that will put the schemas into motion. The solution to changing perceptions is in the framework of schematic organization, such as the values construct, the belief system.

Affect: Despite the thinking processes, people share experiences with others through feelings, emotional reactions and emotional states, which is the affect. People know to link thinking and perceptions to certain sets of emotions which is the consequence of the perception that a certain state of events or facts exists.

Tips for a message to be more effective:

- Is recognized as authority by others.
- Recommends that a benefit offered is uncommon.
- Submits that a benefit has a deadline.
- Is consistent with past behavior or expectations of the audience.
- Appeals to the receiver target group.
- Uses reciprocity by signifying that the sender facilitated in the past and now needs some help in return Chaiken & Maheswaran, 1994).

2.3.3. Listening Skills
People do not listen to everything heard and not everything heard is important to pay much attention to. Listening is an active behavior, a skill that can be

boosted with understanding and practice. The ability to listen is affected by many factors turning people into selective listeners deliberately or unconsciously (Fogg et al, 2002).

Effective communicators are short-term listeners and decision makers listen selectively due to unconscious attitudes about the subject but there are challenging stimuli. A barrier to communication is the variance between the information rate of exhibition by the sender which is usually between 70 to 140 words per minute and the receiver information absorption rate which is much greater speed. This situation in the rate of information transfer creates trouble to communication such as fantasizing, sleeping, frustration and anger (Fogg et al, 2002).

2.3.4. Memory and Retention of Information

In the decision making process Information retention and memory long enough is essential.

Activity: Answer the following questions:
1. What is the name of street where you live?
2. What was your mother's maiden name?
3. What are the names of Santa's reindeer?
4. What was the name of your first-grade school bus driver?

Memory refers to the ability to acquire, retain, and use information and knowledge either short-term memory or long-term memory, so to answer these questions, the answers must have been encoded, stored, and retrieved. Short-term memory is the fragment of memory presently active and can rapidly change when focus of attention is changed and are kept active in the mind for less than ninety seconds.

Long-term memory, on the other hand, is the warehouse all acquired knowledge and information and is retained in the mind for more than an hour. Short-term memory supports the learning process. Working memory is part of the short-term memory that is a speedily disappearing set of information that stores only recent events in the brains.

Sometimes, memory can be misleading by remember information differently than the way they actually occurred because it is complex, and can store information in forms that include events, facts and related environment, perceptions, judgments, and images. People only store information in long-term memory that seems to be significant or valuable for future situation.

The attempt is to persuade the receiver to activate long-term memory through association of new information with information already deposited in the long-term memory, and the store of the new information will be facilitated and repetition of the information by reference, to be relocated to long-term storage space. Reference memory is part of the long-term memory and is all of the unchanging images and facts learned and committed Eagly & Chaiken, 1975). Subjects that follow ones pre-existing beliefs or experience can convert part of his long-term memory subjects that are new or conflict with a people's pre-existing experience.

2.4.   Persuasion and Audience

Persuasion is the communication effort that is anticipated to support, form, and alter the responses, attitudes and behavioral intentions such as decision-making of the audience, and not only communicating objective facts and information. Persuasion to stimulate someone act as he would not act, occurs when a sender of information presents an idea or attitude that differs from ideas or attitudes that are already held by the audience.

The audience may have no predetermined ideas about the subject or may have dissimilar ideas or attitudes about it Giffen & Ehrlich, 1963). The consequence of presentation of a new or different idea is for the audience to feel temporary pressure that produces disagreement, conflict or inappropriateness and have them motivated to attempt to eliminate the disagreement by (1) accepting the new idea willingly, (2) rejecting it, or (3) integrating it with previously held ideas and attitudes (Hogg et al., 1993).

Sometimes persuasion is a challenge to pass new ideas or attitudes are in antagonism or some modification is required to a people's previously beliefs (Cialdini, 1993). The best opportunity when faced with a pre-existing attitude is to find a way to resolve that attitude with the new idea that people wants the audience to accept. The main work is to have an audience research, specially designed to identify the pre-existing attitudes and discover ways to manage acceptance of the message (Byrne, 1971).

2.4.1. Liking

People carelessly tend to agree those they like (Burgoon et al., 2002). Liking as a term, describes the emotional link that an individual may feel toward another person (Smith et al., 2005). Liked senders are more effective stimulus mediators than others (Sampson & Insko, 1964).

Likability is considered to be a persuasion tactic of self-presentation stressed (Kenrick et al.,2002). Enhanced liking for the sender is escorted by boosted conclusions of the communicator's fidelity (O' Keefe, 2002). Further to that, similarity to us increases likability (Carli et al., 1991).

2.5.   The Sender Expertise, Credibility and Authority

The dimensions of credibility are comprised of the elements of trustworthiness and expertise (Fogg 2003; O'Keefe, 2002;).   The sender and the way he behaves are important to credibility and the persuasion this enforces since expertise and fidelity are critical factors in determining credibility. A person with a reputation for expertise and fidelity will generally be perceived as credible.

The expertise of the message sender can serve as a simple signal when the likelihood of thinking is low (Petty, Cacioppo, & Goldman, 1981), can be analyzed as an argument (Kruglanski et al., 2005) and bias the nature of the thoughts (Chaiken & Maheeswaran, 1994) when the likelihood of thinking is high, and can affect the extent of thinking when thinking is unconstrained (De Bono & Harnish, 1988).

People hold the mental shortcut of assuming that others who simply display symbols of authority such as titles and tone should be listened to (Rhoads & Cialdini, 2002). Authority is linked to credibility and persuasiveness (Fogg, 2003).

2.5.1. Three C's of Credibility

Credibility comes from the root word "credo," which means "I believe" (The American Heritage® Dictionary, 2013) and refers to the judgments made by a receiver concerning the believability of a sender (Fogg, Lee and Marshall, 2002; O'Keefe, 2002). The study of credibility in communication began 2500 years ago with Aristotle, who looked at persuasion in debaters and rhetoric was his concept of communication. The modes of audience persuasion are ethos (credibility), logos (logical appeals of the speaker), and pathos (emotional appeals).

Aristotle said that ethos contained three assessments of the speaker by the audience.

- Competence: The speaker knows what they are talking about and has good sense?
- Character: The speaker is reliable, honest and did they have good character?
- Caring: The speaker has the audience's good in mind and has goodwill toward the audience?

It is of great significant to understand that the sender credibility attributed to the source not to the content is deteriorated as time passes, and the message is weakened as the message source is forgotten and distanced from the message (Rhoads & Cialdini, 2002). Contrariwise, the message effect increases over time even though the sender is not observed as credible, expert or of high fidelity. In situations where people of low experience transmit a dominant message, the message is remembered better than the sender we have the development of the sleeper effect.

People favorable to a message (Wells & Petty, 1980) or merchandise (Tom et al., 1991) were asked to nod rather than shake their heads to it. If a message presented weak arguments on an important topic, those nodding their heads reported more confidence in their negative thoughts and thus were less favorable toward it than were those who were shaking their heads (Brinol & Petty, 2003). People perceive expertise when the speaker appears to be knowledgeable and speaks directly and confidently.

It is vital to mention the difference between the terms similarity and credibility (Lascu, et al., 1995). People tend to rely upon the advice of other people who have experienced the same problem rather than an expert but when discuss a more technical topic, they tend to rely upon experts. If the topic relates to a subjective preference (i.e., personal choice), people tend to prefer the opinion of someone who shares their personal values, tastes, or way of life, but when they make judgments about facts or objective reality people prefer the opinion of someone with objective credibility. Sometimes, it is more powerful to use familiar analogies and references rather than expert testimony and opinions learned (McGuire, 1968).

According to balance theory people tend to like others who exhibit signs of similarity because it is reinforcing to their own self-concept and helps them to predict and understand similar others (Heider, 1958), and similarity is the determinants of liking (Michener et al., 2004).

People accept messages from credible and pleasant senders (O'Keefe, 2002) and a more credible sender is more preferred and persuasive (Kelman, 1961; Anderson & Clevenger, 1963). The sender credibility is positively correlated with message recipients' attitude and behavioral intentions and behaviors (Senecal & Nantel, 2004).

People, cognitively choose whom to will trust in which respects and under which circumstances and we base the choice on what we take to be "good reasons," constituting evidence of trustworthiness (Lewis and Weigert, 1985). Trustworthiness is a difficult issue due to bias or motives in making important statements because it can be defines as the personal characteristics that inspire positive expectations on the part of other individuals (Butler & Cantrell, 1984; McKnight et al., 1998).

Only a credible sender can change when supporting an idea that is different from the pre-existing attitude held by the receiver. The more issue-relevant cognitive activity that goes into an attitude change, the more durable the new attitude is and the more impact it has on other judgments and behaviors (Petty, Haugtvedt, & Smith, 1995).

2.6. Brain

Human brain was replaced human muscle power as the engine of firm success (Porat, 1998) and those worked with intangible resources (Drucker, 1959). The brain can be thought as a complex, hierarchical network, in which billions of neurons are organized into circuits, columns, and functional areas where, information processing comes from specific patterns of activity over the neurons, linking brain structure and function which are limited by the structure of brain net (He et al., 2007; Huttenlocher, 2002; Sur and Rubenstein, 2005).

The connection between perception and action is done by a sequence of neural operations, which makes a stimulus to guide behavior to make a decision to choose a particular action or motor response (McGuire, 1968).

2.6.1. Cognitive Activity

Organizations struggle to increase the competitive advantages through continuous innovation and creativity (Mohrman et al., 2002). Sophisticated thinking skills, such as naturalistic decision making, where experts make instinctive decisions based on cognitive skills developed on patterns formed from previous experiences and the decision maker is unable to communicate the process or provide justification for the decision choice and integrative thinking, are required by cognitive work (Kahneman & Klein, 2009). The decision-making process does not include the choice among alternatives

(Lipshitz et al., 2006) but people take options created from their analysis of environmental signals related to experiences (Kahneman & Klein, 2009).

The integrative thinkers take into account irrelevant factors and have the ability to calculate relationships by keeping the parts of the puzzle in their mind simultaneously, and resist simple-minded and distinctive ways of teach (Martin, 2007). Cognitive work requires people to reliably complete un-monotonous tasks, by persistently using new knowledge and abilities (Hackman and Oldham, 1980).

Metacognition refers to these second-order thoughts as about thoughts about our thoughts or thought processes (Petty, Brinol, Tormala, & Wegener, 2007), ant has taken main role in the domain of social psychology (Jost, Kruglanski, & Nelson, 1998), and in memory research (Koriat & Goldsmith, 1996), clinical practice (Beck & Greenberg, 1994), and advertising (Friestad & Wright, 1995).

Self-validation hypothesis is a term that describes that generating thoughts is not sufficient to have an impact on judgments, but must also have confidence in them (Petty, Brinol & Tormala, 2002). It makes some predictions such as:

- Confidence is used in determining which attitudes guide behavior (Fazio & Zanna, 1978).
- Attitude–thought correlations increased as measured thought confidence increased (Petty et al., 2002).
- When positive thoughts had been generated toward the message, experiencing confidence following thought generation led to more persuasion (Brinol, Petty and Barden 2007).
- A message presented prior to receiving an emotion manipulation in which people were required to behave according to a happy or sad script (Velten, 1968).
- Happy people relied more on their thoughts than did sad individuals, but agreement was reduced because the thoughts were unfavorable.

3. Application Suggestions

3.1. Measuring Effectiveness of Communication

There are two ways to measure the effectiveness of communication:

- Evaluate the quality of the communication by comparing the actual communication to its purpose. This measure is an subjective measurement,
- Observe whether the receiver is motivated to behave in accordance with the purpose of the communication. This measure is more objective (Brinol, Petty and Barden 2007).

3.2. Tips for Effective Persuasion

- Know the reality: Researched the evidence is significant for credible sender.
- Know the receiver: Number, types, current opinion, basis for opinion, own needs and interests and arguments to persuade them.
- Express similarities between you and receiver, such as common values, beliefs, and experiences.
- Utilize opinion leaders as more credible a communicator.
- Make a strong opening. When audience attention is at its highest, and when its opinion is the most flexible.
- Get to the point. The audience doesn't have time to waste with long since it drives to lose people's attention, and credibility.
- Offer a benefit supporting your position. Turn these into tailor-made to your audience
- Minimize the costs
- Ask for an action step. Make them participate through an action request.
- The action step should be unambiguous and specific.
- Make the action step simple and feasible.
- Have a selection of action steps available.

- Obtain a commitment to take the step made publicly, to make him more accountable.
- Use models who have taken the desired action, have benefited from it, and are willing to say so publicly that have positive persuasive impact.
- Repeat the message as necessary to make sure that the message and the requested action have fully registered.
- Thank the target person for listening, and for the consideration given.
- Follow up to ensure the committed action has in fact been taken.
- Keep the target person informed.

Regardless of how to evaluate the effectiveness, the central purpose is:
- *Does the communication have its desired effect?*

The communication efforts are often ineffective even though one has communicated with great social skill.

Bibliography

- Anderson, K., & Clevenger, T. (1963). A Summary of Experimental Research in Ethos. Speech Monographs, 30, 59-78.
- Argyris, C. (1977). Double Loop Learning in Organisations. *Harvard Business* Review, Sept.-Oct, pp. 115-124
- Aristotle (translated and organized by Ross, Ross, Ackrill & Urmson), *The Nicomachean Ethics (Oxford World's Classics)* (Oxford: Oxford University Press, 1998).
- Aristotle (translated by Barnes), *The Complete Works of Aristotle,* Vol. 2 (Princeton: Princeton University Press, 1984).
- Asch, S.E. (1948). The doctrine of suggestion, prestige, and imitation in social psychology. Psychological Review, 55, 250–276.
- Barney, J. (1991). Firm resources and sustained competitive advantage. Journal of Management, 17(1): 99– 120.
- Baumard, Ph. (1994b). From Noticing to Making Sense: Using Intelligence to Develop Strategy. Journal of Intelligence and Counterintelligence, pp. 7.
- Beck, A.T., & Greenberg, R.L. (1994). Brief cognitive therapies. In A.E. Bergin & S.L. Garfield (Eds.), Handbook of psychotherapy and behavior change (pp. 230–249). New York: Wiley.
- Beer, M. (1980). Organization Change and Development: a Systems View, Goodyear, Santa Monica, CA.
- Bovee, C. L. and Thill, J.V. (1992). Business Communication Today, (3rd Edition), McGraw-Hill, Inc.
- Brin⁻ ol, P., Petty, R.E., & Barden, J. (2007). Happiness versus sadness as a determinant of thought confidence in persuasion: A self-validation analysis. Journal of Personality and Social Psychology, 93, 711–727.
- Brin⁻ ol, P., Petty, R.E., Valle, C., Rucker, D.D., & Becerra, A. (2007). The effects of message recipients' power before and after persuasion: A self-validation analysis. Journal of Personality and Social Psychology, 93, 1040–1053.

- Brinol, P., & Petty, R.E. (2003). Overt head movements and persuasion: A self-validation analysis. Journal of Personality and Social Psychology, 84, 1123–1139.

- Brinol, P., Petty, R.E., & Tormala, Z.L. (2004). The self-validation of cognitive responses to advertisements. Journal of Consumer Research, 30, 559–573.

- Brinol, P., Petty, R.E., Gallardo, I., & DeMarree, K.G. (2007). The effects of self-affirmation in non-threatening persuasion domains: Timing affects the process. Personality and Social Psychology Bulletin, 33, 1533–1546.

- Brown, AW and Kaka, A. (2003). Project management strategic issues. Edinburgh: Herriot Watt University.

- Brownstein, A. L. (2003). Biased predecision processing. Psychological

- Bulletin, 129(4), 545–569.

- Burgoon, J. K., Dunbar, N. E. & Segring, C. (2002). Nonverbal Influence. In Dillard, J. P. & Pfau, M. (Eds), Persuasion Handbook: Developments in Theory and Practice , pp.445-473. Thousand Oaks, CA: Sage Publications.

- Burke W. (2002), Organization Change: Theory and Practice, SAGE Publications

- Butler, J. K., Jr., & Cantrell, R. S. (1984). A behavioral decision theory approach to modeling dyadic trust in superiors and subordinates.

- Byrne, D. (1971). The attraction paradigm. New York: Academic Press.

- Calfee, J. E., & Ringold, D. J. (1994). The 70% majority: Enduring consumer beliefs about advertising. Journal of Public Policy and Marketing, 13, 228–238.

- Carli, L. L., Ganley, R., & Pierce-Otay, A. (1991). Similarity and satisfaction in roommate relationships. Personality and Social Psychology Bulletin , 17, 419-426.

- Chaiken, S., & Maheswaran, D. (1994). Heuristic processing can bias systematic processing: Effects of source credibility, argument ambiguity, and task importance on attitude judgment. Journal of Personality and Social Psychology, 66, 460–473.

- Cialdini, R. B. (1993). *Influence: Science and practice* (3rd ed.). New York : HarperCollins.

- DeBono, K.G., & Harnish, R.J. (1988). Source expertise, source attractiveness, and processing or persuasive information: A functional approach. Journal of Personality and Social Psychology, 55, 541–546.

- DeSteno, D., Petty, R.E., Wegener, D.T., & Rucker, D.D. (2000). Beyond valence in the perception of likelihood: The role of emotion specificity. Journal of Personality and Social Psychology, 78, 397–416.

- DeVito, J.A.(1986). The communication handbook: A dictionary. New York: Harper & Row. The American Heritage® Dictionary of the English Language, 5th edition Copyright © 2013 by Houghton Mifflin Harcourt Publishing Company. Published by Houghton Mifflin Harcourt Publishing Company.

- Drucker, P. (2001). The next society. Economist, 361(8246), 3-5. Equal Employment Opportunity Commission, U. S. Civil Service Commission, U. S. Department of Labor & U. S. Department of Justice. (1978). Uniform guidelines on employment selection procedures. Federal Register, 43, No. 166, 38290-38309.

- Drucker, P. F. (1959). Landmarks of tomorrow. New York: Harper & Row.

- Drucker, P. F. (1999) Knowledge worker productivity: The biggest challenge. California Management Review, 41(2), 79-94.

- Eagly, A. H., & Chaiken, S. (1975). An attribution analysis of the effect of communicator characteristics on opinion change: The case of communicator attractiveness, Journal of Personality and Social Psychology , 32, 136-144.

- Fazio, R.H., & Williams, C.J. (1986). Attitude accessibility as a moderator of the attitude-perception and attitude-behavior relations: An investigation of the 1984 presidential election. Journal of Personality and Social Psychology, 51, 505–514.

- Fazio, R.H., & Zanna, M.P. (1978). Attitudinal qualities relating to the strength of the attitude-behavior relationship. Journal of Experimental Social Psychology, 14, 398–408.

- Fogg, B. J., Lee, E., & Marshall, J. (2002). Interactive technology and Persuasion. Dillard, J. P., & Pfau, M. (Eds). Persuasion handbook: Developments in theory and practice. (pp.765-797). London: United Kingdom.

- Fogg. B.J., (2003). Persuasive Technology: Using Computers to Change What We Think and Do. San Francisco, CA: Morgan Kaufmann Publishers.

- Foucault, M. (1992). The Order Of Discourse, Brutus Ostlings Bokforlag Symposion (Stockholm/Stehag).

- Friestad, M., & Wright, P. (1995). Persuasion knowledge: Lay people's and researches' beliefs about the psychology of persuasion. Journal of Consumer Research, 27, 123–156.

- Giffen, K., & Ehrlich, L. (1963). Attitudinal effects of a group discussion on a proposed change in company policy. Speech Monographs , 30, 377-379.

- Gupta, V., Hanges P. J. and Dorfman P. W. (2002). Cultural Clusters: Methodology and Findings, Journal of World Business, Volume 37:1, 11 - 15.

- Hackman, J. R., & Oldham, G. R. (1980). Motivation through the design of work. In J. R.

- Hall, P. A., and Taylor, R. C. R. (1996). Political Science and the Three New Institutionalisms. Political studies, 44 (5), pp. 936-957.

- Heider, F. (1958). The psychology of interpersonal relations . New York: John Wiley.

- Hodgers, M. R. and Luthans, F. (1991). International Management, McGraw-Hill, Inc.

- Hofstede, G. (1983). Dimensions of National Cultures in Fifty Countries and Three Regions, In J. Deregowski, S. Dziurawiec, and R.C. Annis (Eds.), Expectations in Cross-Cultural Psychology. Lisse, Netherlands: Swets and Zeitilinge.

- Hogg, M. A., Cooper-Shaw, L., & Holzworth, D. W. (1993). Group prototypicality and depersonalized attraction in small interactive groups, Personality and Social Psychology Bulletin , 19, 452-565.
- Huttenlocher, P. R. & Dabholkar, A. S. (1997). Development of the Prefrontal Cortex: Evolution, Neurobiology, and Behavior (eds Krasnegor, N. A., Lyon, G. R. & Goldman-Rakic, P. S.) 69–84 (Paul. H. Brookes, Baltimore).
- Jost, J.T., Kruglanski, A.W., & Nelson, T.O. (1998). Social meta-cognition: An expansionist review. Personality and Social Psychology Review, 2, 137–154.
- Kahneman, D., & Klein, G. (2009). Conditions for intuitive expertise: A failure to disagree. American Psychologist, 64(6), 515-526.
- Kegan, D. L., & Rubenstein, A. H. (1973). Trust, effectiveness, and organizational development: A field study in R&D. The Journal of Applied Behavioral Science, 9, 498–513.
- Kelly J. R. and Bowles, G. (2006). Value and task management. Edinburgh: Herriot Watt University.
- Kelman, H. C. (1961). Processes of opinion change. Public Opinion Quarterly, 25, 57-78.
- Kenrick, D. T., Neuberg, S. L., & Cialdini, R. B. (2002). Social psychology: Unraveling the mystery (2nd ed.). Boston: Allyn & Bacon.
- Kline, R. B. (2005). Principles and practices of structural equation modeling (2nd ed.). New York: Guilford Press.
- Koriat, A., & Goldsmith, M. (1996). Monitoring and control processes in the strategic regulation of memory accuracy. Psychological Review, 103, 490–517.
- Krosnick (Eds.), Attitude strength: Antecedents and consequences (pp. 93–130). Mahwah, NJ: Erlbaum.
- Kruglanski, A.W., Raviv, A., Bar-Tal, D., Raviv, A., Sharvit, K., Ellis, S., et al. (2005). Says who? Epistemic authority effects in social judgment. In M.P. Zanna (Ed.), Advances in experimental social psychology (Vol. 37, pp. 346–392). San Diego: Academic Press.

- Kuchinke, K. P. (1999). Leadership and culture: work-related values and leadership styles among one company's U.S. and German telecommunication employees', Human Resource Development Quarterly 10(2): 135–54.

- Lascu, D.-N., Bearden, W. O., & Rose, R. L. (1995). Norm extremity and personal influence on consumer conformity. Journal of Business Research, 32(3), 201-213.

- Lewis, J. D., & Weigert, A. (1985). Trust as a social reality. Social Forces, 63, 967–985.

- Lipshitz, R., Klein, G., & Carroll, J. S. (2006). Introduction to the special issue. Naturalistic decision making and organizational decision making: Exploring the intersections. Organization Studies, 27(7), 917-923.

- Luthans, F. and. Hodgers, M. R. (1989). Business Communication. The Dryden Press.

- Martin, L.L., Ward, D.W., Achee, J.W., & Wyer, R.S. (1993). Mood as input: People have to interpret the motivational implications of their moods. Journal of Personality and Social Psychology, 64, 317–326.

- Martin, R. (2007). How successful leaders think. Harvard Business Review, 85(6), 60-67.

- McCormick, E. J. (1979). Job analysis: Methods and applications. New York, NY: AMACOM.

- McGuire, W. J. (1968). The Nature of Attitudes and Attitude Change. In Lindzey, G. and Aronson, E. (Eds.). Handbook of Social Psychology . MA: Addison-Wesley.

- McKnight, D. H., Cummings, L. L., & Chervany, N. L. (1998). Initial trust formation in new organizational relationships. Academy of Management Review, 23, 473–490.

- Michener, H. A., DeLamater, J. D., & Myers, D. J. (2004). Social Psychology (5th edition).CA.

- Mohrman, S. A., Finegold, D., & Klein, J. A. (2002). Designing the knowledge enterprise: Beyond programs and tools. Organizational Dynamics, 31(2), 134-150.

- O'Keefe, D. J. (2002). Persuasion: Theory & Research . Thousand Oaks, CA: Sage Publications.
- Perloff, (1993)*The Dynamics of Persuasion* (Mahwah, NJ: Lawrence Erlbaum Associates,.
- Petty, R. E., Wegener, D. T., & Fabrigar, L. R. (1997). *Attitudes and attitude change*. Annual Review of Psychology, 48, 609-647.
- Petty, R.E., & Cacioppo, J.T. (1979). Issue involvement can increase or decrease persuasion by enhancing message-relevant cognitive responses. Journal of Personality and Social Psychology, 37, 1915–1926.
- Petty, R.E., & Cacioppo, J.T. (1981). Attitudes and persuasion: Classic and contemporary approaches. Dubuque, IA: William C. Brown.
- Petty, R.E., & Cacioppo, J.T. (1984). The effects of involvement on responses to argument quantity and quality: Central and peripheral routes to persuasion. Journal of Personality and Social Psychology, 46, 69–81.
- Petty, R.E., & Cacioppo, J.T. (1986). Communication and persuasion: Central and peripheral routes to attitude change. New York: Springer-Verlag.
- Petty, R.E., & Krosnick, J.A. (Eds.). (1995). Attitude strength: Antecedents and consequences. Mahwah, NJ: Erlbaum.
- Petty, R.E., & Wegener, D.T. (1998). Attitude change: Multiple roles for persuasion variables. In D. Gilbert, S. Fiske, & G. Lindzey (Eds.), The handbook of social psychology (4th ed., Vol. 1, pp.323–390). New York: McGraw-Hill.
- Petty, R.E., &Wegener, D.T. (1999). The elaboration likelihood model: Current status and controversies. In S. Chaiken & Y. Trope (Eds.), Dual process theories in social psychology (pp. 41–72). New York: Guilford Press.
- Petty, R.E., Brin˜ ol, P., Tormala, Z.L., & Wegener, D.T. (2007). The role of meta-cognition in social judgment. In E.T. Higgins & A.W. Kruglanski (Eds.), Social psychology: A handbook of basic principles (2nd ed., pp. 254–284). New York: Guilford Press.

- Petty, R.E., Cacioppo, J.T., & Goldman, R. (1981). Personal involvement as a determinant of argument-based persuasion. Journal of Personality and Social Psychology, 41, 847–855.

- Petty, R.E., Cacioppo, J.T., & Heesacker, M. (1981). The use of rhetorical questions in persuasion: A cognitive response analysis. Journal of Personality and Social Psychology, 40, 432– 440.

- Petty, R.E., Fabrigar, L.R., & Wegener, D.T. (2003). Emotional factors in attitudes and persuasion. In R.J. Davidson, K.R. Scherer, & H.H. Goldsmith (Eds.), Handbook of affective sciences (pp. 752– 772). Oxford, England: Oxford University Press.

- Petty, R.E., Haugtvedt, C., & Smith, S.M. (1995). Elaboration as a determinant of attitude strength: Creating attitudes that are persistent, resistant, and predictive of behavior. In R.E. Petty & J.A.

- Petty, R.E., Ostrom, T.M., & Brock, T.C. (1981). Cognitive responses in persuasion. Hillsdale, NJ: Erlbaum.

- Petty, R.E., Schumann, D.W., Richman, S.A., & Strathman, A.J. (1993). Positive mood and persuasion: Different roles for affect under high and low elaboration conditions. Journal of Personality and Social Psychology, 64, 5–20.

- Petty, R.E., Wheeler, S.C., & Bizer, G.Y. (1999). Is there one persuasion process or more? Lumping versus splitting in attitude change theories. Psychological Inquiry, 10, 156–163.

- Petty, R.E.,Wells, G.L., & Brock, T.C. (1976). Distraction can enhance or reduce yielding to propaganda: Thought disruption versus effort justification. Journal of Personality and Social Psychology, 34, 874– 884.

- Porat, M. U. (1998). The information economy: Definition and measurement. In J. W. Cortada (Ed.), Rise of the knowledge worker (pp.101-132). Boston, MA: Butterworth-Heinemann.

- Psychological Reports, 55, 19–28.

- Rhoads, K. V. & Cialdini, R. B. (2002). The Business of Influence. In Dillard, J. P. & Pfau, M. (Eds), Persuasion Handbook: Developments in

Theory and Practice (pp. 513-542). Thousand Oaks, CA: Sage Publications.

- Russo, J. E., Medvec, V. H., & Meloy, M. G. (1996). The distortion of information during decisions. Organizational Behavior and Human Decision Processes, 66, 102−110.

- Sampson, E. E., & Insko, C.A. (1964). Cognitive consistency and performance in the autokinetic situation. Journal of Abnormal and Social Psychology, 68, 184-192.

- Schwarz, N., Bless, H., & Bohner, G. (1991). Mood and persuasion: Affective status influence the processing of persuasive communications. In M. Zanna (Ed.), Advances in experimental social psychology (Vol. 24, pp. 161−197). San Diego, CA: Academia Press.

- Senecal, S. & Nantel, J. (2004). The influence of online product recommendations on consumers' online choices. Journal of Retailing, 80, 159-169.

- Sifianou, M. (2001). Discourse Analysis: an Introduction, Leader Books.

- Smith, D., Menon, S., & Sivakumar, K. (2005). Online Peer and Editorial Recommendations, Trust, and Choice in Virtual Markets. Journal of Interactive Marketing, 19(3), 15-37.

- Smith, E.R , & DeCoster, J. (2000). Dual process models in social and cognitive psychology: Conceptual integration and links to underlying memory systems. Personality and Social Psychology Review, 4, 108−131.

- Thompson (1999), *Persuading Aristotle: The Timeless Art of Persuasion in Business, Negotiation, and the Media* (St. Leonards, Australia: Allen & Unwin,.

- Tiedens, L.Z., & Linton, S. (2001). Judgment under emotional certainty and uncertainty: The effects of specific emotions on information processing. Journal of Personality and Social Psychology, 81, 973−988.

- Tom, G., Pettersen, P., Lau, T., Burton, T., & Cook, J. (1991). The role of overt head movement in the formation of affect. Basic and Applied Social Psychology, 12, 281–289.
- Tormala, Z.L., Brin˜ ol, P., & Petty, R.E. (2006). When credibility attacks: The reverse impact of source credibility on persuasion. Journal of Experimental Social Psychology, 42, 684–691.
- Tormala, Z.L., Petty, R.E., & Brin˜ ol, P. (2002). Ease of retrieval effects in persuasion: The roles of elaboration and thought-confidence. Personality and Social Psychology Bulletin, 28, 1700–1712.
- Velten, E. (1968). A laboratory task for induction of mood states. Behavior Research and Therapy, 6, 473–482. 146 Volume 3—Number 2 Persuasion Processes
- Wegener, D.T., & Petty, R.E. (1997). The flexible correction model:The role of naive theories of bias in bias correction. In M.P. Zanna (Ed.), Advances in experimental social psychology (Vol. 29, pp. 141–208). San Diego: Academic Press.
- Wells, G.L., & Petty, R.E. (1980). The effects of overt head movements on persuasion: Compatibility and incompatibility of responses. Basic and Applied Social Psychology, 1, 219–230.

About the author:

Prof. Dimitrios P. Kamsaris

Prof. Dimitrios P. Kamsaris is the Chairman of the Academic Board and Vice President at Bilston Community College in the UK, responsible for the International Affairs. He is a Professor of Organizational Behavior in Switzerland and Visiting Professor of Management and Marketing at Business Schools in UK, Denmark, France, Cyprus, and Greece

Prof. Kamsaris has completed postdoctoral education at Harvard University. He held CEO and managerial positions at: Coca-Cola, Sherwin Williams, Olympic Games of 2004, Shell and D Constructions.

Today, he is a strategic consultant in major companies and trainer of public & private sector executives in the U.K., Denmark, KSA, UAE, Qatar, Oman, Cyprus, Greece and Pakistan.

He is published in numerous business, academic journals and books.